Zombies in Nature

BY KIRSTEN W. LARSON

AMICUS HIGH INTEREST ❦ AMICUS INK

Amicus High Interest and Amicus Ink are imprints of Amicus
P.O. Box 1329, Mankato, MN 56002
www.amicuspublishing.us

Library of Congress Cataloging-in-Publication Data
Larson, Kirsten W., author.
 Zombies in nature / Kirsten W. Larson.
    pages cm. – (Freaky nature) (Amicus high interest)
 Summary: "This photo-illustrated book for elementary readers
describes animals that seem to take over the brains of other
animals. Explains how parasites can alter their hosts' behaviors
to use the host for their survival"– Provided by publisher.
 Audience: K to grade 3.
 Includes bibliographical references and index.
 ISBN 978-1-60753-783-0 (library binding)
 ISBN 978-1-60753-882-0 (ebook)
 ISBN 978-1-68152-034-6 (paperback
 1. Parasites–Behavior–Juvenile literature. 2. Parasitism–Juvenile
literature. 3. Animal behavior–Juvenile literature. I. Title.
 QL757.L37 2016
 577.8'57–dc23

                                            2014038738

Editor: Wendy Dieker
Series Designer: Kathleen Petelinsek
Book Designer: Heather Dreisbach
Photo Researcher: Derek Brown

Photo Credits: Julian Money-Kyrle / Alamy cover; Photoshot
Holdings Ltd./Alamy 5; Cavallini James/BSIP/Superstock
6; Marcel Derweduwen/Shutterstock 8–9; shunfa Teh/
Shutterstock 11; DP Wildlife Invertebrates/Alamy 12; Pascal
Goetgheluck/Science Source 15; Thomas Hahmann/
Wikimedia 16; Alberto Carrera/age fotostock/Superstock 19;
Jane Leaman/Alamy 20; FLPA/Superstock 22–23; Prof. José
Lino-Neto/Wikimedia 24; Science Photo Library/Alamy 27;
NHPA/Superstock 28

Printed in Malaysia

HC 10 9 8 7 6 5 4 3 2 1
PB 10 9 8 7 6 5 4 3 2 1

Table of Contents

What Is a Zombie?

In stories and movies, zombies are dead people. But they have come back to life. A magic spell controls them. They are like slaves to someone else. Most of these tales are made up. But some animals act like zombies. Yikes! Animals called **parasites** control them. Parasites are living things that feast on others. Their victims are called **hosts**.

This fish louse is a parasite.
It doesn't make fish turn
into zombies, though.

Roundwoms lay their tiny eggs in a host's belly. We need a microscope to see them.

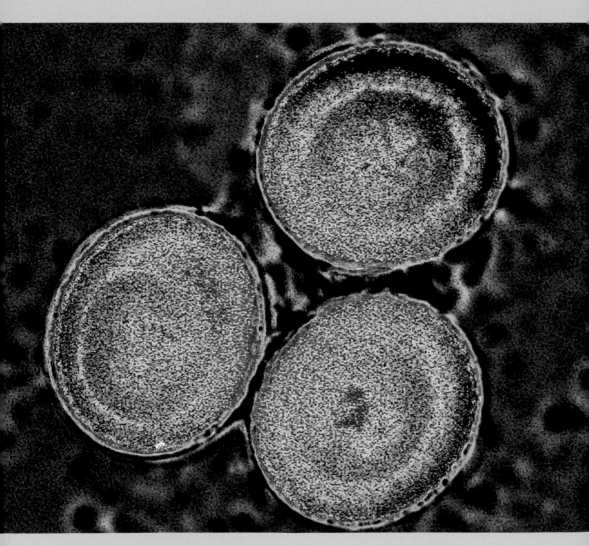

 Where does the word zombie come from?

Sometimes, parasites live in an animal's belly. They eat some of the victim's food. Other times the parasite or its babies eat the host itself. They might slurp the host's body fluids, like blood. Gross! But zombie-making parasites are special. They turn their hosts into slaves. They force them to do strange things. The hosts act like zombies.

It comes from Haiti. It means "spirit of the dead." Stories in Haiti tell of a witch that raises the dead and makes them slaves.

Is that rat a zombie? Normally, a rat hides. It knows cats might be near. But it is not afraid if a parasite **infects** its brain. Now the rat is a zombie! It stays out in the open. And the cat pounces. Gulp! The parasite has a new home—inside the cat! It comes out in the cat's poop. Then it can find another host.

A rodent that doesn't hide from danger might be a zombie.

Zombie-Making Fungus

Carpenter ants live high in rainforest trees. But a **fungus** attacks an ant's brain. The ant acts like a zombie. It goes to the ground where it is cool and wet. That is where the fungus grows best. Soon the ant dies. A stalk grows from the ant's head. It spreads **spores**. Spores are like seeds that make more fungus. Watch out, ants!

A fungus sprouts from an ant's head. But first it forces the ant to carry it to the forest floor.

This fly's legs are covered in white zombie-making fungus.

 **Q** How does the fly get stuck up high?

Sometimes a fungus sends spores down from up above. One fungus attacks houseflies. First a spore bores into the fly. Inside, it makes more spores. The fly's body puffs up. It cracks open. The fly is a zombie! It climbs up high and sticks itself to a window or leaf. Then a shower of spores rains down. Next victim!

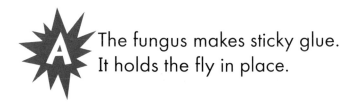 The fungus makes sticky glue. It holds the fly in place.

Wiggly Worms

In a river, a newly hatched worm waits. A baby bug gobbles it up. That bug lives a normal life. But when it dies, things get weird. A cricket eats the dead bug and becomes a zombie. Normally, crickets hate water. But the horsehair worm makes the cricket dive in. Splash! A clump of worms crawls from the cricket.

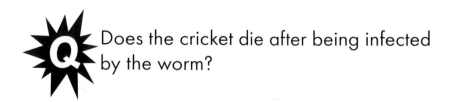

Q Does the cricket die after being infected by the worm?

A horsehair worm bursts out of a zombie cricket.

 A Usually it does. It drowns. Sometimes a cricket climbs from the water and lives.

The snail's colorful eyestalk is actually a worm. It has taken control of this snail!

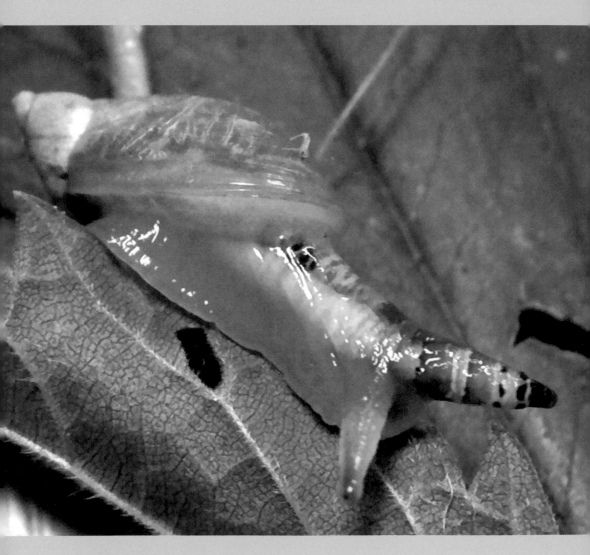

 Why would the worm trick the bird?

Sometimes parasites are tricky. An unsuspecting snail slurps bird poop with worm eggs in it. Gross! When the eggs hatch, the zombie-making starts. The worm crawls up the snail's eyestalk. It grows fat and colorful. Then the worm forces the snail to climb to a sunny leaf. A bird thinks the snail's eye is a caterpillar. Chomp!

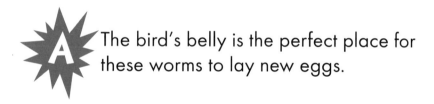
The bird's belly is the perfect place for these worms to lay new eggs.

The roundworm is another tricky zombie-maker. An ant eats bird poop filled with worm eggs. Now the ant is a zombie! The worms make the ant stay out in the open. They also make the ant's **abdomen** look red like a berry. A bird is fooled and eats the ant. Gulp! The worms lay eggs in the bird's belly. Soon more ants will be zombies.

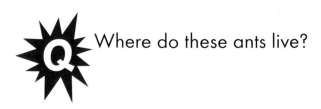 Where do these ants live?

Rainforest birds sometimes eat zombie ants filled with worms.

 The rainforests of Central and South America.

19

A strong spiderweb with a zombie spider can be a good place for a wasp cocoon.

 **Q** Why does the wasp need the web?

Wild Wasps

A mother wasp lays her egg on a spider. When the egg hatches, the baby wasp stings the spider. Ouch! The spider is now under the baby's control. The wasp makes the spider spin a strong web. The web will hold the wasp's **cocoon**. When the spider is finished, the wasp eats it. How is that for thanks?

 It keeps the cocoon from getting washed away in heavy rains.

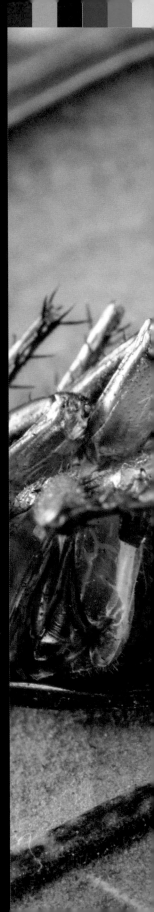

A mother jewel wasp stings a cockroach. It is   . Then she stings the helpless roach's brain. Zap! The roach becomes a zombie. Using the roach's antenna, the wasp leads the roach to her nest. There she lays an egg in the roach's body. When the wasp baby hatches, it eats the roach. Dinnertime!

A wasp comes out of a cockroach's body. It hatched inside and ate its way out.

23

This zombie caterpillar is
protecting some wasp cocoons.

 Q What happens after the wasps leave
their cocoons?

Another wasp turns caterpillars into zombie babysitters. A mother wasp lays her eggs inside the caterpillar. When the babies hatch, they eat their way out. Then they spin cocoons. Meanwhile, the caterpillar stands guard. If predators come close, the zombie scares them off. How does the wasp control the caterpillar? Scientists still don't know.

 The caterpillar dies. Its job is done.

Zombies in the World

Most people don't believe in zombies. We don't see dead people walking around. There are not slaves under a magic spell. But in nature, parasites do make slaves of their hosts. Why? Sometimes just for a meal. Hosts often are food for the zombie-maker. They would die without the hosts.

Very small parasites like these
need a host to live in to survive.

Making slaves of hosts also helps parasites find new hosts. They can find hosts in new places. But only with the help of other animals. Sometimes hosts even care for parasite babies. But things often do not end well for the zombies. Most die when their jobs are done. And zombie-makers live to see another day.

A fungus sprouts out all over this ant. Soon another ant will become a zombie!

Glossary

abdomen The back part of an insect's body.

cocoon A case made by an insect where it stays while changing into an adult.

fungus A plant-like living thing that has no leaves, flowers, or roots.

host An animal that a parasite uses to survive.

infect To spread throughout a part of the body; parasites might infect a host's brain.

parasite An animal that lives off of another; parasites might eat a host's food inside its belly, drink its host's blood, or eat it after using the host's body for some other purpose.

spore A seed-like part of a fungus that produces a new fungus.

stunned To be unable to move or think; frozen.

Read More

Davies, Nicola. *What's Eating You? Parasites— The Inside Story.* Cambridge, Mass.: Candlewick Press, 2009.

Johnson, Rebecca L. *Zombie Makers: True Stories of Nature's Undead.* Minneapolis: Millbrook Press, 2013.

Nagle, Frances. *Zombie Caterpillars.* New York: Gareth Stevens, 2016.

Websites

Neuroscience for Kids—How Do Parasites Hijack Their Host's Brains?
https://faculty.washington.edu/chudler/toxo.html

So You Want a Pet Parasite?—Discovery Kids
discoverykids.com/games/so-you-want-a-pet-parasite/

UC Berkeley News—Ant Parasite Discovered
www.berkeley.edu/news/media/releases/2008/01/ants-vid.shtml

Index

About the Author

Kirsten W. Larson used to work with rocket scientists at NASA. Now she writes about science for children. She's written more than a dozen articles for children's magazines, as well as six books. Zombies do not scare her one bit. She lives with her husband and two boys in California.